Plastic

by Melanie Mitchell

Lerner Publications Company · Minneapolis

Look at all of the **plastic**.

There are many kinds and colors of plastic.

Plastic is made in **factories**.

This factory makes plastic parts for cars.

Radios are made of plastic.

Computers have plastic parts, too.

Kids play with plastic toys.

Plastic toys are fun at the **beach**.

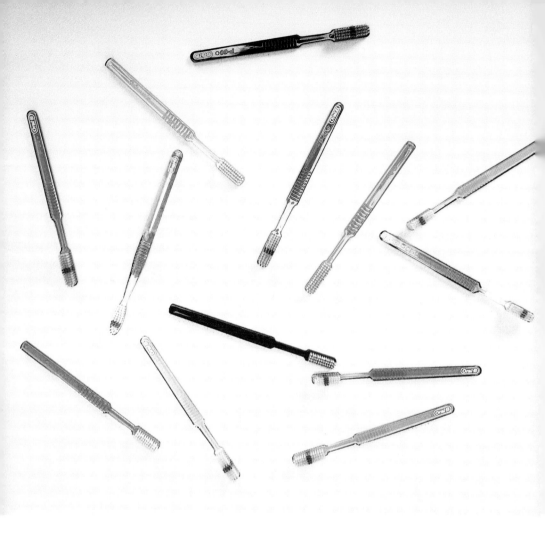

Plastic toothbrushes clean our teeth.

We drink out of plastic cups.

Milk comes in plastic bottles.

Lots of food comes in plastic, too.

People throw away a lot of plastic.

We can **recycle** plastic.

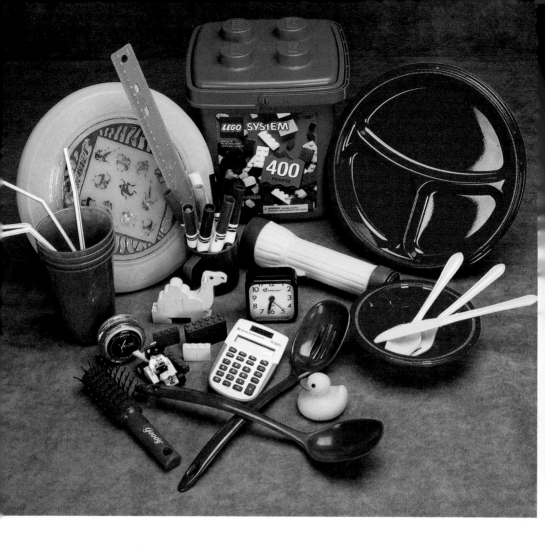

Many things are made out of plastic.

Plastic is important to us.

Recycling Plastic

Look around you. What do you see that is made of plastic? People use lots and lots of plastic. When plastic is thrown away, it goes to a landfill. Plastic can be in landfills for hundreds of years. Most plastic can be recycled. When plastic is recycled, it is melted and used to make other things. Recycled plastic can be used to make carpet, flower pots, grocery bags, and many other things. See if you can recycle some plastic today.

Plastic Facts

 Plastic was first invented in 1860 by Alexander Parkes.

 Plastic soda bottles can be recycled to make carpeting and clothing.

 It takes 24 gallons of water to make 1 pound of plastic.

 Over 80 million tons of plastic are produced in the world each year.

Americans use more than 2.5 million plastic bottles every hour.

 Plastic sandwich bags were first sold in 1957.

 Most of Australia's money feels and looks like paper, but it is really made of thin plastic.

Glossary

 beach – the sandy shore of an ocean, lake, or river

 computers – machines that store information and help do work

 factories – places where things are made or put together

 plastic – a material made from chemicals

 recycle – change something that has been thrown away so it can be used again

Index

bottles – 12

factories – 4

kinds and colors – 3

recycle – 15

toothbrushes – 10

toys – 8, 9

Copyright © 2003 by Lerner Publications Company

All rights reserved. International copyright secured. No part of this book may be reproduced, stored in a retrieval system, or transmitted in any form or by any means—electronic, mechanical, photocopying, recording, or otherwise—without the prior written permission of Lerner Publications Company, except for the inclusion of brief quotations in an acknowledged review.

The photographs in this book are reproduced through the courtesy of: © Todd Strand/Independent Picture Service, front cover, pp. 4, 6, 7, 8, 10, 11, 13, 16, 22 (second from top, middle); PhotoDisc, pp. 2, 12, 14, 22 (second from bottom); © Wolfgang Kaehler/CORBIS, p. 3; © Jim West, p. 5; Eyewire, pp. 9, 15, 22 (top, bottom); © Richard Cummins, p. 17.

Illustration on page 19 by Laura Westlund.

Lerner Publications Company
A division of Lerner Publishing Group
241 First Avenue North
Minneapolis, MN 55401 U.S.A.

Website address: www.lernerbooks.com

Library of Congress Cataloging-in-Publication Data

Mitchell, Melanie S.
    Plastic / by Melanie Mitchell.
       p.   cm. — (First step nonfiction)
    Includes index.
    Summary: An introduction to plastic and its uses in everyday life.
    ISBN: 0-8225-4620-5 (lib. bdg. : alk. paper)
    1. Plastics—Juvenile literature. [1. Plastics.] I. Title. II. Series.
TP1125 .M58 2003
620.1'923—dc21   2002006473

Manufactured in the United States of America
1 2 3 4 5 6 – JR – 08 07 06 05 04 03